MAP SKILLS, G

TABLE OF CONTENTS

Introduction 1
Glossary 2
Pretest 3
Posttest 4
United States Map 5
World Map 6

Unit 1: Map Basics
Directions 7
Making a Compass 8
Cardinal Directions 9
Intermediate Directions 10
Where Is It? 11
Parts of a Map 12
Using a Map Grid 13
Making a Map Index 14
Where Are You Now? 15

Unit 2: How to Measure Distance
Using a Distance Scale 16
Farther Than It Looks 17
Trip Planner 18

Unit 3: How to Use a Route Map
Route Maps 19
Getting Around Town 20
Are We There Yet? 21
Using a Mileage Table 22

Unit 4: How to Find Capitals and Boundaries
State Capitals and Boundaries 23
National Capitals and Boundaries 24
Who's in Charge? 25

Unit 5: How to Use a Landform Map
Landform Maps 26
Relief Maps 27
Elevation Maps 28
Rivers 29

Unit 6: How to Use a Resource Map
Resource and Product Maps 30
Land Use Maps 31
A Useful Map 32

Unit 7: How to Use Latitude and Longitude
Latitude 33
Longitude 34
Latitude and Longitude 35
Can You Find It? 36

Unit 8: How to Use Globes and World Maps
Globes 37
World Maps 38
Hemispheres 39
The Eastern Hemisphere 40
North America 41

Unit 9: How to Use a Historical Map
Historical Maps 42
Trails to the Past 43

Unit 10: How to Use Special Purpose Maps
Time Zone Maps 44
Population Maps 45
Precipitation Maps 46
Movement Maps 47

Answer Key 48

INTRODUCTION

The world is constantly changing. Current events carry us to places around the globe. For today's students to be well informed, they must know the basic skills of geography and map use. Map skills help students to improve their sense of location, place, and movement. By giving students a better knowledge of geography and maps, we give them a better understanding of the world in which they live. Such knowledge will help students in their standardized testing and in their broader academic pursuits. And some people need to know how to use a street map just to get across town! *Map Skills* is meant to address these needs and many more.

General standards in geography for this grade suggest that students should be able to perform a variety of skills. Students should be able to understand and use map components, employ the cardinal and intermediate directions, use latitude and longitude, identify the continents and oceans, and recognize landforms. *Map Skills* provides students with extensive practice in these areas.

Organization
The book is divided into ten units, each centering on a particular map component or type of map. Many of the pages also contain additional activities that put the map skills into practical use. The book also contains a pretest and posttest to assess students' strengths and weaknesses in using maps.

Maps and Globes
Students gain a greater sense of place by knowing their relationship to other places. For this reason, maps and globes are important tools. The book contains a varied group of maps for student use. One or two pages recommend the use of a globe, so if one is available, it should be prominently displayed in the classroom. In any case, various maps, especially of the United States and the world, are a handy addition to any classroom.

Glossary

boundary (p. 23) the dividing line on a map where one place ends and another place begins. A boundary is also sometimes called a **border** (p. 23).

cardinal directions (p. 9) the four main directions of north, south, east, and west

compass (p. 8) a device for determining directions

compass rose (p. 9) a symbol that shows the directions on a map

continent (p. 37) a very large body of land. There are seven continents.

degree (p. 33) the unit of measurement used for lines of latitude and longitude

direction (p. 7) the line or course along which something is moving or pointing

distance (p. 16) how far one place is from another, often measured in miles or kilometers

distance scale (p. 16) the guide to what the distances on a map stand for

elevation map (p. 28) a map that shows the elevation, or height, of the land's surface

Equator (p. 33) the imaginary line that goes around the middle of the Earth. The Equator divides the Earth into the Northern and Southern Hemispheres.

globe (p. 37) a spherical model of the Earth

grid (p. 13) a pattern of lines that cross each other to form squares or rectangles

hemisphere (p. 39) half of the globe or half of the Earth. The four hemispheres are Northern, Southern, Eastern, and Western.

historical map (p. 42) a map that shows a place during another time in history

inset map (p. 43) a small map within a larger map

intermediate directions (p. 10) the in-between directions of northeast, southeast, southwest, and northwest

intersection (p. 10) the place where two or more routes meet or cross

landform map (p. 26) a map that shows the shape of the land, such as mountains and hills. A landform map is also known as a **physical map** (p. 29) or a **relief map** (p. 27).

legend (p. 12) another name for a map key

lines of latitude (p. 33) lines that circle the Earth north and south of the Equator. They are numbered and marked by degrees.

lines of longitude (p. 34) lines that circle the Earth from the North Pole to the South Pole. They are numbered and marked by degrees.

location (p. 7) tells where something can be found

map (p. 9) a drawing of a real place. A map shows the place from above.

map index (p. 14) the alphabetical list of places on a map identified by their grid section

map key (p. 12) the guide to what the pictures or symbols on a map stand for

mileage (p. 21) distance in miles

mileage table (p. 22) a table on a map that tells how far in miles one place is from another

national capital (p. 24) a place where laws are made for a nation or country

ocean (p. 37) a very large body of water. There are four oceans.

population map (p. 45) a map that shows the density of population in a certain area

precipitation map (p. 46) a map that shows how much precipitation falls in a place

Prime Meridian (p. 34) the line of longitude from the South Pole to the North Pole measured at 0°. It divides the Earth into the Eastern and Western Hemispheres.

resource (p. 30) something people can use, such as oil, trees, and water

resource map (p. 30) a map that shows the location of resources in a place. A resource map is also sometimes known as a **product map** (p. 30) or a **land use map** (p. 31), which shows how people make use of the land itself.

route (p. 19) a road or path from one place to another. Highways, railroads, and trails are routes.

route map (p. 19) a map that shows the locations and intersections of routes. Route maps are also called **road maps** or **street maps** (p. 19).

sea level (p. 28) the height of 0 feet on an elevation map

state capital (p. 23) the center of government in a state, where laws are made for that state

symbols (p. 12) drawings or patterns that show what things on a map mean

time zone (p. 44) an area of the Earth where the time is the same. The Earth has 24 time zones.

title (p. 12) a name that identifies a map and its contents

Name _____ Date _____

Pretest

Directions  Study the map of Ashland. Then, darken the circle by the correct answer to each question.

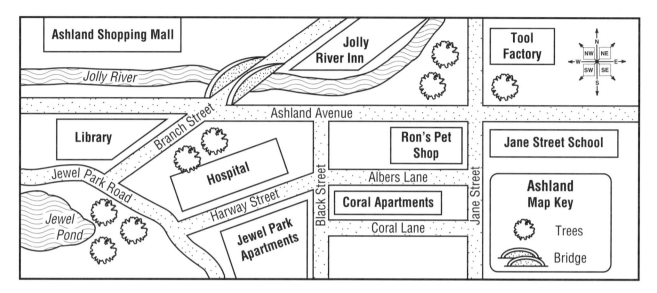

1. What does 🌳 stand for on the map?
   - Ⓐ bridge
   - Ⓑ house
   - Ⓒ trees
   - Ⓓ birds

2. Ron's Pet Shop is _____ of the Coral Apartments.
   - Ⓐ north
   - Ⓑ south
   - Ⓒ east
   - Ⓓ west

3. The Jewel Park Apartments are _____ of the hospital.
   - Ⓐ northwest
   - Ⓑ southwest
   - Ⓒ southeast
   - Ⓓ northeast

4. The _____ is northeast of Ron's Pet Shop.
   - Ⓐ Jolly River Inn
   - Ⓑ Coral Apartments
   - Ⓒ Jane Street School
   - Ⓓ Tool Factory

5. Jane lives in the Coral Apartments. She wants to go to the Ashland Shopping Mall. Which direction must she go?
   - Ⓐ northwest
   - Ⓑ southwest
   - Ⓒ southeast
   - Ⓓ northeast

6. Branch Street runs _____.
   - Ⓐ northwest and southeast
   - Ⓑ northwest and northeast
   - Ⓒ northeast and southwest
   - Ⓓ northwest and southwest

Name _____ Date _____

Posttest

Directions → Study the map. Use the map to answer the questions.

WESTERN EUROPE

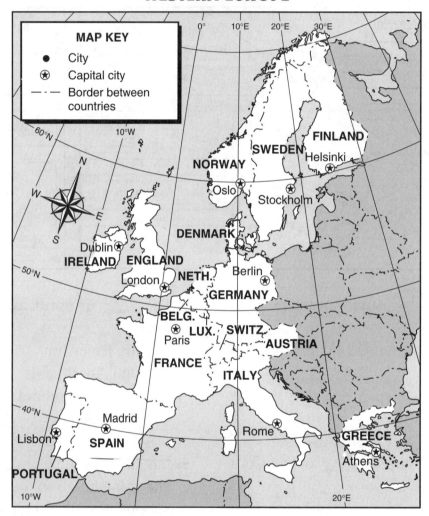

1. What is the national capital of Italy? _____

2. What is the national capital of Germany? _____

3. You travel from Greece to Ireland. Which direction do you go? _____

4. What city is located at 60° N, 10° E? _____

5. Which city is located on the Prime Meridian? _____

6. What are the latitude and longitude of Helsinki, Finland? _____

Name _____ Date _____

United States Map

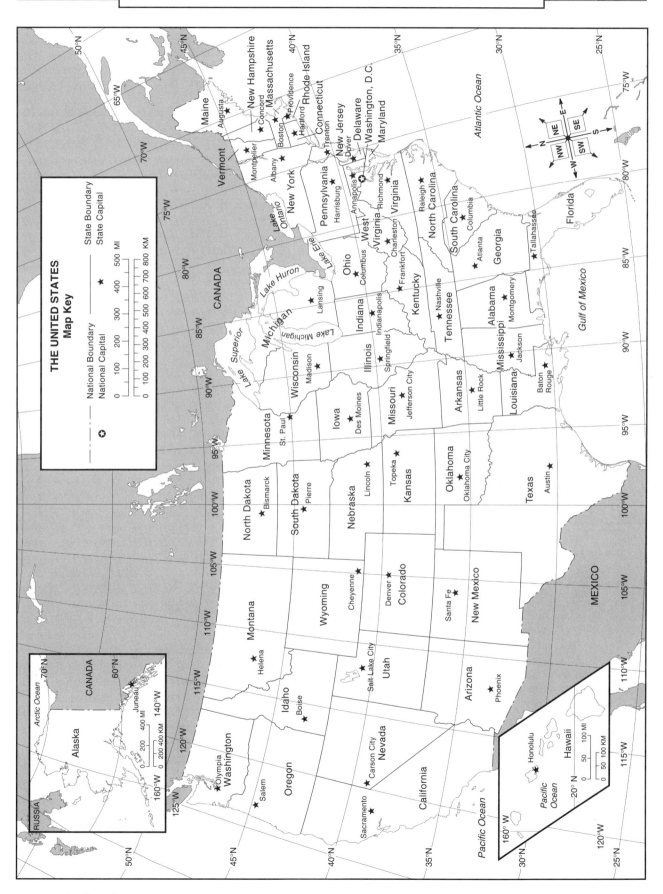

www.svschoolsupply.com
© Steck-Vaughn Company

United States Map
Map Skills 4, SV 6132-8

Name _____ Date _____

World Map

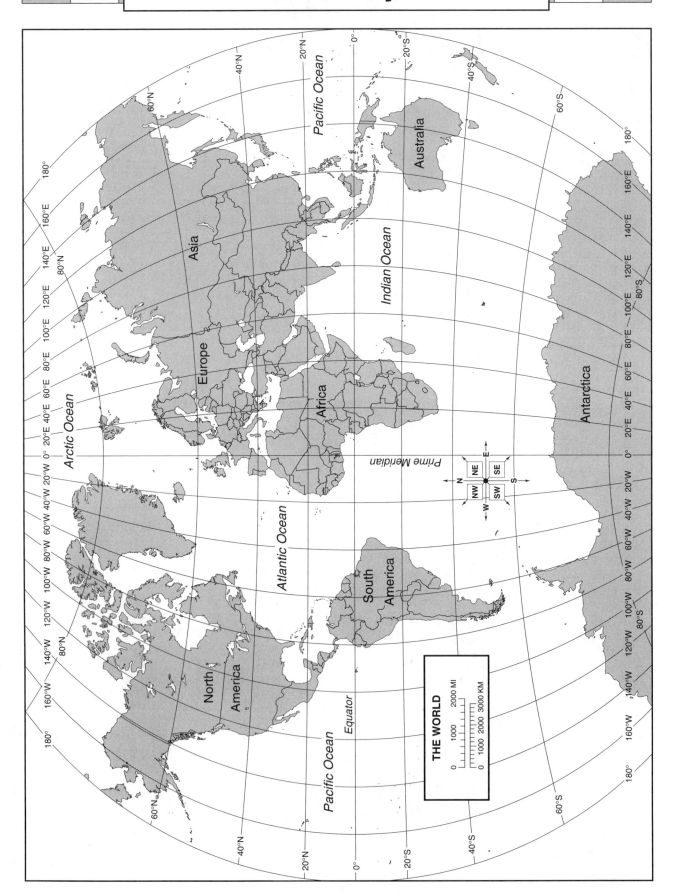

www.svschoolsupply.com
© Steck-Vaughn Company

World Map
Map Skills 4, SV 6132-8

Directions

You use **directions** every day. Directions help you to compare **locations**. Directions tell you where you are in relation to other places. They also tell you where other places are in relation to you. You might say:

"The chalkboard is in the front of the room."
"Go up the stairs to get to the office."
"Turn left at the lockers to reach the library."
"I'll wait outside the cafeteria."

You might combine directions: "The bookcase is to my right when I face the front of the room." But you could also say: "The bookcase is to my left when I face the back of the room." Both sets of directions tell about the same place.

Activity 1

Draw a "map" of your classroom. Label the front of the room front. Also label the back. Draw in the chalkboard, the door, and the windows. Also draw in the desks, bookcases, and computer station. Add any other features you want. When you finish, share your map with the class. See how your map differs from the maps of your classmates. Save your map for use in another activity later.

Activity 2

Choose a place in your classroom. Give a classmate directions to that place. You might say, "This place is on the right of the room when you face the front." After each wrong guess, give more directions. Count the guesses until your partner is correct. Then, switch places. Let your partner choose a place and give directions. See who can guess a place in the fewest tries. Repeat the activity for three places in the room.

Name _____   Date _____

Making a Compass

A **compass** is an instrument with a magnetized needle. A compass is used to find the directions of north, south, east, and west. The needle of the compass always points north. If you know where north is, you can find all the other directions. You can make your own compass. Here's how.

You will need:
1 sewing needle (It's sharp, so be careful.)
1 magnet (such as one on the refrigerator)
1 small plastic bowl of water
1 small piece of paper, about 2 inches square
1 real compass

Steps:

1. Stroke the **dull** end of the needle across the magnet. Do this about 60 times. Be sure to stroke the same direction each time.

2. Put the bowl of water on a table. The bowl should not be made of metal. Make sure there is no metal within 2 feet of the bowl. The table should not be metal. Check to see there is no metal under the table. Also, make sure the magnet you used is not nearby. The real compass should not be close, either.

3. Now, set the paper on top of the water in the center of the bowl. Set the needle on top of the paper. Move the paper to make it spin slightly. If the paper gets stuck on the side of the bowl, move it back to the center.

4. Watch the needle. What is happening? The needle and paper should stop moving completely. When they do, the **sharp** end of the needle should be pointing north.

5. Check how accurate your homemade compass is. Use the real compass. Don't get the two compasses too close to each other. If you do, they will interfere with one another.

6. Now, you know which direction is north. Face north. South will be behind you. East will be to your right. West will be to your left.

Activity

Pretend that you are lost in the woods. It is 3 P.M., and the Sun is shining brightly. But you are worried that you might not get out of the woods before dark. You know that you have to go south to get back to the campground. It would be easy to find your way out with a compass. But you forgot the compass in your tent. How will you get out of the woods? Write a paragraph telling what you would do.

Name _____   Date _____

Cardinal Directions

    A **map** is a drawing of a real place. It shows the place as if you are looking from above. A map uses the directions of north, south, east, and west. These are called the **cardinal directions**. Look on the map for a direction marker. This marker is called a **compass rose**. It may have the abbreviations for all the cardinal directions: *N, S, E, W*. Or it may have only the abbreviation for north: *N*. If it has only the *N*, you have to remember how to find the other directions. You can use the memory tip: Never Eat Shredded Wheat.

    Face a wall map. Find north. It should be toward the top of the map. South is the opposite of north, so it should be toward the bottom of the map. East is toward the right side of the map. West is toward the left side.

    One thing to remember is that north <u>is not</u> up. North <u>is</u> at the top of the map. But pretend that the map is lying flat on the ground. Now the compass rose points along the ground. North is a direction along the ground, not up in the air.

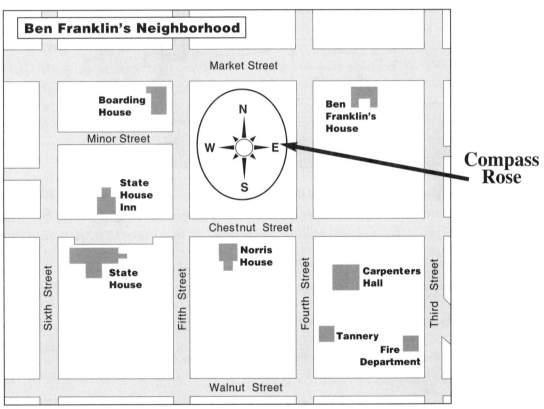

Activity

Look at the map. Is Market Street north or south of Chestnut Street? Let's find out. Put your finger on Chestnut Street. Now, look at the compass rose. Remember, north is toward the top of the map. South is toward the bottom of the map. Move your finger toward Market Street. Does your finger move toward the top or the bottom of the map? It moves toward the top. That means Market Street is north of Chestnut Street.

Name _____ Date _____

Intermediate Directions

The compass rose on a map shows the four main directions. These are the cardinal directions: north, south, east, west. Some maps also include in-between directions. These are the **intermediate directions**: northeast, southeast, southwest, northwest. They fall between the cardinal directions. The intermediate directions are abbreviated *NE, SE, SW, NW.*

Look at the map below. Find the compass rose. Find southwest. It falls between south and west. Practice the other intermediate directions, too.

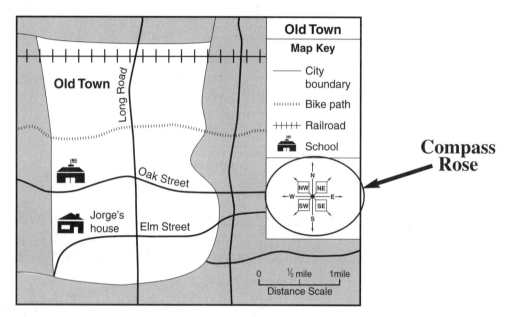

Directions → Look at the map of Old Town. Study the compass rose. Notice the intermediate directions. They are abbreviated *NE, SE, SW, NW.* Move your finger in each intermediate direction. Then, use the compass rose to answer the questions.

1. Find where Long Road crosses Oak Street. This is called an **intersection**. Is this intersection southwest or northeast of Jorge's house? _____

2. Jorge leaves his house and walks southwest. Will he cross Long Road? _____

✦ Activity

The compass rose on a map helps you to find directions. Sometimes, though, it is hard to use. You can make a compass rose that moves. It will be easier to use. Get a small piece of clear plastic about 2 inches square. On it, draw your own compass rose with a black marker. Put the four cardinal directions on the four points. Add the intermediate directions, too. Now, you can put the compass rose anywhere on the map. Try it out on the map above. Remember, north points toward the top of the map.

www.svschoolsupply.com
© Steck-Vaughn Company

10

Unit 1: Map Basics
Map Skills 4, SV 6132-8

Name _____   Date _____

Where Is It?

Intermediate directions give you more ways to describe a place. You could say, "The mall is north of the school." Or you could say, "The mall is northwest of the school." The second description gives a better location. Practice describing locations using intermediate directions.

Sometimes the compass rose will not give the abbreviations for the intermediate directions. The compass rose may have only small lines pointing in the intermediate directions. Then, you will have to remember what those directions are. Look at the map below. Find the compass rose. It does not have the abbreviations for the intermediate directions. Write them in.

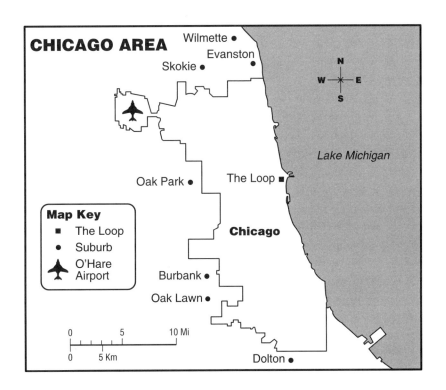

Directions ▶ Study the map of the Chicago area. Then, darken the circle by the answer that correctly completes each sentence.

1. Oak Park is _____ of The Loop.
   - Ⓐ north
   - Ⓑ south
   - Ⓒ east
   - Ⓓ west

2. _____ is southeast of Burbank.
   - Ⓐ Evanston
   - Ⓑ Dolton
   - Ⓒ Oak Park
   - Ⓓ The Loop

3. Skokie is _____ of Wilmette.
   - Ⓐ northwest
   - Ⓑ southwest
   - Ⓒ southeast
   - Ⓓ northeast

4. The Loop is _____ of Oak Lawn.
   - Ⓐ southeast
   - Ⓑ northwest
   - Ⓒ northeast
   - Ⓓ southwest

Name _____ Date _____

Parts of a Map

Some maps have many parts. Some maps have only a few parts. In both cases, these parts help you to know what the map is about. You already know about the compass rose. A map usually has a **title**. The title tells you what information the map contains. Many maps also have a **map key**, or **legend**. Maps may use **symbols**, drawings or patterns to show what things on the map mean. The map key tells you what these symbols mean.

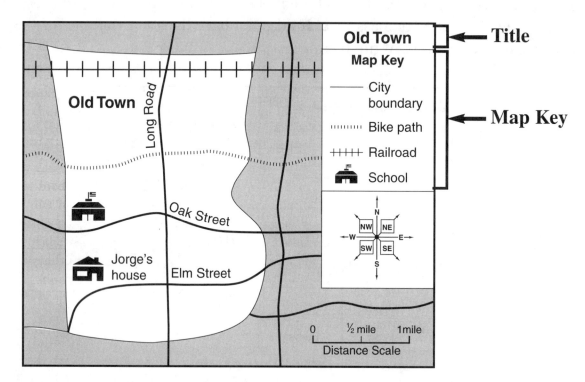

Directions → Look at the map of Old Town. Find the title and map key. Study these parts carefully. Then, answer the questions.

1. What is the title of this map? _____
2. What does this symbol 🏫 in the map key stand for? _____
3. What does this symbol ++++ in the map key stand for? _____
4. What does this symbol ······ in the map key stand for? _____

☼ **Activity**

Use your classroom map from page 7. Have your teacher help you to find north. Then, add a compass rose to your map. Make north the top of your map. Add the title "My Classroom." Also add a map key that uses symbols for the desks, bookcases, and computer station.

www.svschoolsupply.com
© Steck-Vaughn Company

12

Unit 1: Map Basics
Map Skills 4, SV 6132-8

Name _____   Date _____

Using a Map Grid

Some maps have a grid to help you to locate places. A map **grid** is a pattern of lines that cross each other. The lines form squares or rectangles. Each square or rectangle is a grid section. Each row of squares or rectangles has a letter. Each column of squares or rectangles has a number.

Look at the map below. The letters are on the left side: A, B, C, D. The numbers are on the top: 1, 2, 3. Each rectangle on the map can be named with a letter and a number.

Here is how to find grid section C-2. Put your finger on the letter **C.** Move your finger along the row. Stop when you get to column **2**. That grid section is named C-2. What is in grid section C-2? The fire station is in that section.

Directions ▶ Use the map to answer the questions.

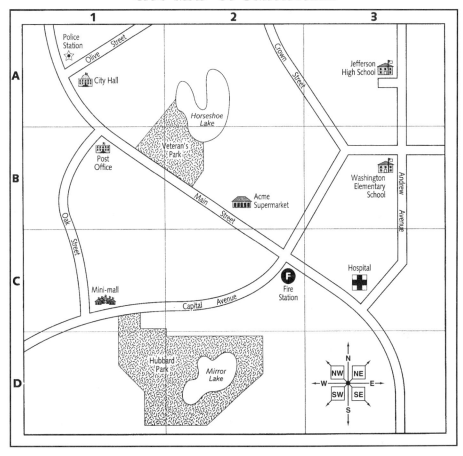

CITY MAP OF SMITHVILLE

1. In which grid section is the post office? _____

2. What place is in grid section C-3? _____

3. In which two grid sections are schools located? _____

4. In which three grid sections is Hubbard Park located? _____

☼ **Activity**

Use your classroom map from page 12. Draw a grid on your map. Put letters on the left side. Put numbers at the top. In which grid section is your desk?

www.svschoolsupply.com
© Steck-Vaughn Company

Unit 1: Map Basics
Map Skills 4, SV 6132-8

Name _____   Date _____

Making a Map Index

A **map index** is an alphabetical list of places on a map. The map index tells you in which grid section a place is located. Suppose you need to find the zoo in Sacramento. Look at the words that start with **Z** in the map index below. Read the grid section letter and number by the word <u>Zoo</u>. The zoo is in grid section C-2.

Now, find the zoo on the map. Put your finger on the **C** row. Move your finger across to column **2**. You have found the zoo.

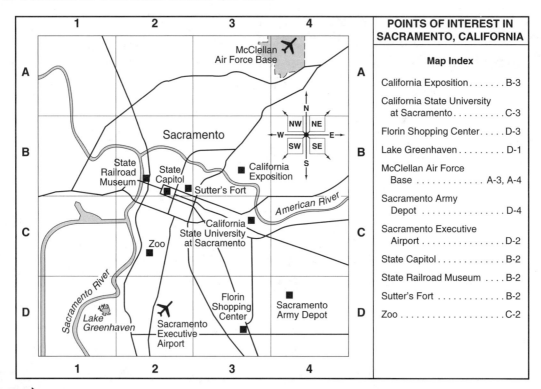

Directions ▶ Study the map and the map index. Use them to answer the questions.

1. In which grid section is the State Capitol located? _____

2. You are at Lake Greenhaven. In which grid section of the map are you? _____

   You want to travel from the lake to the zoo.

   Which direction must you go? _____

3. In which grid section can you find a shopping center? _____

4. In which grid section can you find a fort? _____

Activity

Use your classroom map from page 13. Use the grid to make a map index. Include these terms: Teacher's desk, Bookcase, My desk, Computer station. Put the terms in alphabetical order. Write the grid square letter and number by each term.

Name _____   Date _____

Where Are You Now?

Now you know many of the things necessary to use a map. Let's see how good your map skills are so far.

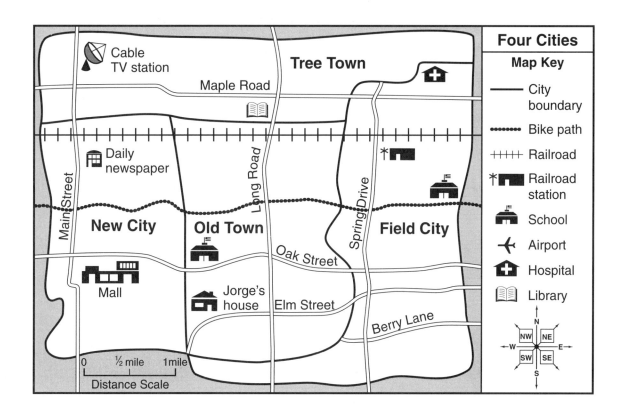

Directions Study the map of the four cities. Then, use all you have learned so far to finish each sentence.

1. The title of this map is _____.
2. Old Town is _____ of Field City.
3. The railroad station is _____ of Jorge's house.
4. Jorge's house is _____ of the hospital.
5. The hospital is _____ of the mall.
6. The cable TV station is _____ of the library.
7. Long Road runs _____ and _____.
8. The railroad track runs _____ and _____.

Name _____ Date _____

Using a Distance Scale

Maps are not life-size. They show big areas on small pieces of paper. **Distance** on a map is not the same as the real distance in a place. To show distance, maps use a **distance scale**. A distance scale shows that a certain length on the map equals a certain length in a real place. Most distance scales show distance in miles and kilometers.

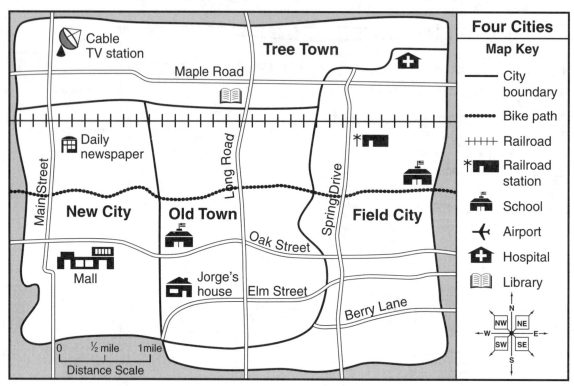

Find the distance scale on the Four Cities map. It is in the lower left corner. On this map, 1 inch on the map equals 1 mile in the real place. Suppose you want to find the distance from the library to the cable TV station. You can do this different ways. You can use the edge of a piece of paper. Lay the paper in a straight line between the two places. Mark the paper below each place. Lay the paper on the scale. Your left mark should be at 0. Your right mark goes beyond the end of the scale. Make a mark on your paper at the 1-mile mark on the distance scale. This mark equals 1 mile. Move this mark to 0. Now, your right mark should be on 1 mile. So, the distance is 1 mile + 1 mile = 2 miles.

Another way is to use an inch ruler. Measure the distance between the library and the cable TV station. The distance is 2 inches. You know from the distance scale that 1 inch = 1 mile. So, 2 inches = 2 miles.

Directions ➤ Use the map to answer the questions.

1. About how far is Jorge's house from the mall? _____

2. About how far is the mall from the library? _____

Name _____ Date _____

Farther Than It Looks

Some maps show a very large place. Then, an inch on the map may equal hundreds of miles. Look at the map below. On this distance scale, 1 inch = 300 miles. Notice that this map also shows distances in kilometers.

Directions Use an inch ruler and the distance scale on the map to finish each sentence.

1. On the map, the distance from Alexandria to Cairo is about _____ inch.

2. On the map, 1 inch = 300 miles. In the real place, Alexandria is about _____ miles from Cairo.

3. On the map, 1 inch = 480 kilometers. In the real place, Alexandria is about _____ kilometers from Cairo.

4. On the map, the distance from Aswan to Cairo is about _____ inches.

5. In the real place, Aswan is about _____ miles from Cairo.

6. Asyut is a city about 250 miles northwest of Aswan. Asyut is about 200 miles south of Cairo on the Nile River. Place the city of Asyut on your map. Draw a dot and write the name <u>Asyut</u>.

Activity

Use your classroom map from page 14. How accurate is it? Draw a distance scale for your map. Let 1 inch = 4 feet. Measure distances in your classroom. Are the distances on your map to scale? If not, redraw the parts of your map that are not accurate.

Name _____   Date _____

Trip Planner

Knowing real distances helps you to plan trips or activities better. For example, you know that a place is a mile away. So, you know about how much time it will take to get there. If you have to travel 25 miles, you have a good idea how long the trip will take.

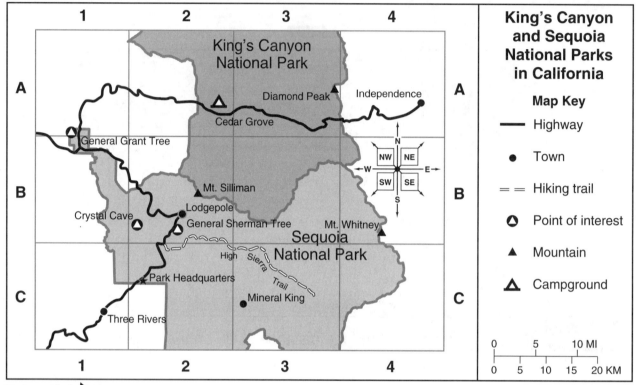

Directions Use the map to answer the questions.

1. In which grid squares are the following places?

Independence _____   Mineral King _____   Cedar Grove _____

Park Headquarters _____   Lodgepole _____   Mt. Whitney _____

2. High Sierra Trail goes through which 3 squares? _____

Directions Complete the table. Use the compass rose to find the direction. Use the distance scale to find the distance.

		Direction	Miles
3.	from Lodgepole to Diamond Peak		
4.	from General Sherman Tree to General Grant Tree		
5.	from Crystal Cave to Park Headquarters		
6.	from Independence to Cedar Grove		

Name _____   Date _____

Route Maps

A **route** is a way to get from one place to another. A route might be a path, a street, a road, or a highway. It could also be a railroad track or even a river. Most people are familiar with **route maps**. They are also called **road maps** or **street maps**. These maps show you how to use streets, roads, or highways to get from one place to another.

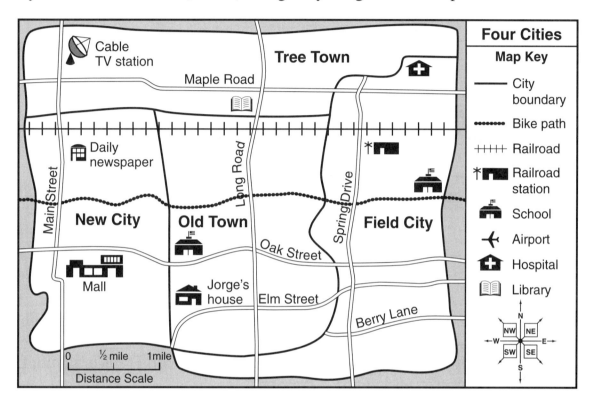

Look at the map of the four cities. Suppose Jorge is at his house. He needs to go to the hospital in Field City. How would you tell him to go? You would name streets he should follow. Use your finger to trace the route you would use to go from Jorge's house to the hospital. Now, write the route he should follow. Include directions.

_____

_____

_____

Did you find a way for Jorge? You might say, "Go east on Elm Street. Turn north on Spring Drive. Then, turn east on Maple Road."

Is there only one route for Jorge to go to the hospital? No, there are several routes. Practice finding the other routes Jorge might use. Write them on another piece of paper.

☼ Activity

Find a route from Jorge's house to some other place on the map. Then, write the set of directions. Give the directions to a classmate. Can the classmate find the place?

www.svschoolsupply.com
© Steck-Vaughn Company

19

Unit 3: How to Use a Route Map
Map Skills 4, SV 6132-8

Name _____ Date _____

Getting Around Town

Many people use street maps to get around their city. The street maps show all the streets in the city. A person can see how to go from one place to another. Sometimes, the person must make many turns and go on many streets to travel somewhere.

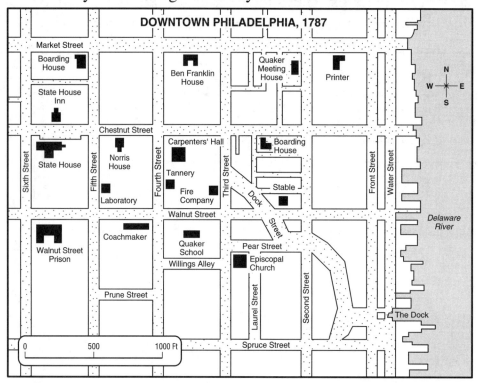

Directions Study the map of downtown Philadelphia in Ben Franklin's time. Use the map to answer the questions.

1. What route could Ben Franklin use to get from his house to the laboratory? Include street names and directions. _____
   _____

2. About how far would Franklin have to walk from his home to the laboratory?
   _____

3. Which four streets go east toward the Delaware River? _____

4. Start at the State House Inn. Go east on Chestnut Street. Turn south on Third Street. Turn east on Walnut Street. Go just past Dock Street.

   Where are you? _____

Activity

Find a route from Ben Franklin's house to some other place on the map. Then, write the set of directions. Give the directions to a classmate. Can the classmate find the place?

Name _____   Date _____

Are We There Yet?

When people go on trips, they often use maps to help them find their way. A good route map makes the travel easier. So learning how to read a route map is an important skill. Many route maps show highways. There are several kinds of highways. There are interstate highways, United States (or U.S.) highways, and state highways. Each kind of highway has a different kind of sign.

Study the map key. It shows the different kinds of highways.

Some route maps do not use a distance scale. Instead, they have numbers to show distance in miles, or **mileage**. Look at the Jones County map. See the **9** between Collier and Jerzy? That number means the distance between the two places is 9 miles. Let's say you need to drive from Jerzy to Collier and then on to Bullock. How far is the trip? From Jerzy to Collier is 9 miles. From Collier to Bullock is 8 miles. Add the mileage. The trip is 17 miles.

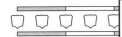

Directions → Study the map. Use the map to answer the questions.

1. What highway runs from O'Reilly to Russell? _____

2. Which highways are shown crossing the river? _____

3. How far is Lesford from O'Reilly? _____

4. What would be the <u>shortest</u> route from Bullock to Schnell? _____

☼ Activity

Tera and her mom live in Bullock. Tera's mom wants to go on an afternoon drive. She wants to drive about 90 miles round trip. Plan a route for the afternoon drive. Include highway numbers, directions, and distances.

Name _____   Date _____

Using a Mileage Table

Some route maps include a **mileage table**. A mileage table tells you how far one place is from another place. Study the mileage table. Four city names are listed on the side. Those same names are at the top. To find how many miles Billings is from Great Falls, find <u>Billings</u> on the side. Run your finger to the right until you reach the <u>Great Falls</u> column. The distance is 224 miles.

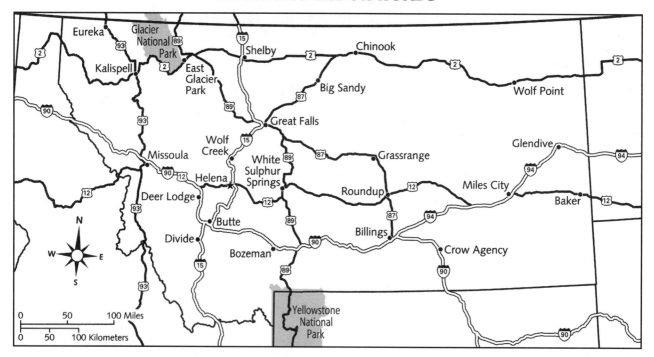

Directions → Study the map and the mileage table. Use the mileage table to determine the distance for each of these trips. Then, trace each trip on the route map using a different color.

1. Butte to Billings _____
2. Billings to Glendive _____
3. Great Falls to Butte _____
4. Glendive to Great Falls _____

	Billings	Butte	Glendive	Great Falls
Billings		236	221	224
Butte	236		457	157
Glendive	221	457		353
Great Falls	224	157	353	

www.svschoolsupply.com
© Steck-Vaughn Company

22

Unit 3: How to Use a Route Map
Map Skills 4, SV 6132-8

Name _____ Date _____

 ## State Capitals and Boundaries

The United States is divided into 50 states. Each state has a center of government. That center is called the **state capital**. Look at the map key below. A star stands for the state capital. Can you find the state capital of Connecticut?

This map also shows the **boundaries** of Connecticut. Boundaries are the lines on a map that show where one state ends and another state begins. The boundaries separate the two states. Boundaries are also called **borders**. Find the state boundary symbol in the map key. Move your finger along the state boundary.

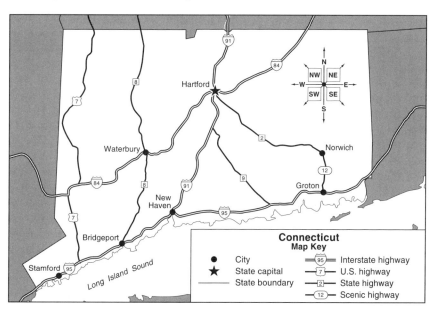

Directions ➤ Study the map. Use the map to answer the questions.

1. What is the capital of Connecticut? _____

2. To go from Waterbury to the state capital, which direction must you travel?
   _____

3. What body of water is on the southern boundary of Connecticut?
   _____

4. What route would you use to go from Bridgeport to the state capital?
   Include highway names and directions. _____
   _____

 Activity

A state's border gives the state its shape. Some states have unusual shapes. Wyoming is a rectangle. The lower part of Michigan looks like a mitten. Find your state on the United States map on page 5. What shape does your state have? Move your finger along the state border.

Name _____   Date _____

National Capitals and Boundaries

Countries are also called nations. Each nation has a center of government. That center is called the **national capital**. Laws for the nation are made in the national capital. In the map key, a star inside a circle stands for a national capital.

Each nation also has a boundary or border. A national boundary separates one nation from other nations. Usually, the places separated by boundaries on a map are different colors or shades.

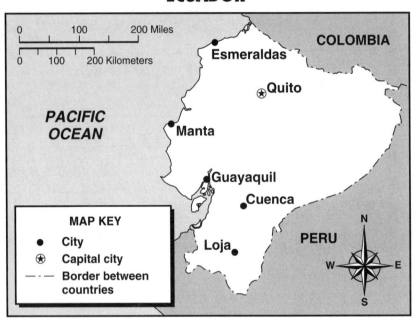

Directions ▶ Study the map. Use the map to answer the questions.

1. Ecuador is a nation in South America. What is the national capital of Ecuador?

   _____

2. What nation is southeast of Ecuador? _____

3. About how far is Manta from the national capital of Ecuador?

   Miles _____   Kilometers _____

4. What body of water is on the western boundary of Ecuador? _____

5. You want to travel from Esmeraldas to the national capital of Ecuador. How many miles and which direction must you go?

   Miles _____   Direction _____

Activity

Find Peru and Colombia on a world map, a globe, or a map of South America. What are the national capitals of Peru and Colombia?

Name _____   Date _____

Who's in Charge?

Now you know how to find capitals and boundaries. Let's see just how much you've learned.

 **Directions** Study the map of the United States on page 5. Use the map to answer the questions.

1. What is the state capital of Ohio?
_____

2. What is the national capital of the United States?
_____

3. What is the state capital of Wyoming?
_____

4. What is the state capital of Georgia?
_____

5. What nation is north of Montana?
_____

6. What nation is south of Texas?
_____

7. You want to travel from the state capital of North Dakota to the state capital of West Virginia. Which direction do you go?
_____

8. You want to travel from the state capital of Arizona to the state capital of Alabama. Which direction do you go?
_____

9. What is the state capital of the state directly south of Kentucky?
_____

10. What is the state capital of the state directly north of New Mexico?
_____

Name _____ Date _____

Landform Maps

A **landform map** tells about the form, or shape, of the land. It tells where mountains and hills are. It tells where plains and deserts are. It can also tell where rivers are.

Alaska is the biggest state in the United States. Alaska has three major landforms: plains, hills, and mountains. Look at the map of Alaska. Study the map key. On this map, mountains are shown by little pointed lines. Dots stand for plains. Gray shading stands for hills. Other maps may use different symbols for the different landforms.

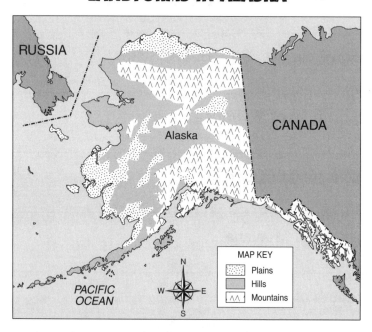

Directions Study the map. Use the map to answer the questions.

1. What kind of landform is found along the northern coast of Alaska?

   _____

2. What kind of landform is mostly found in the southeast part of Alaska?

   _____

3. What kind of landform is found on the islands in southwest Alaska?

   _____

4. Are there more plains or mountains in Alaska?

   _____

Activity

What kinds of landforms are in your state? Are there mountains or hills? Are there plains or deserts? Do research to learn about the landforms in your state. Write a short report on what you learn.

Relief Maps

Sometimes a landform map is called a **relief map**. This kind of relief means the uneven quality of a land surface. A mountain is higher than a hill, and a hill is higher than a plain. So, the heights of these landforms are not even. In this map, the hilly areas around mountains are called highlands.

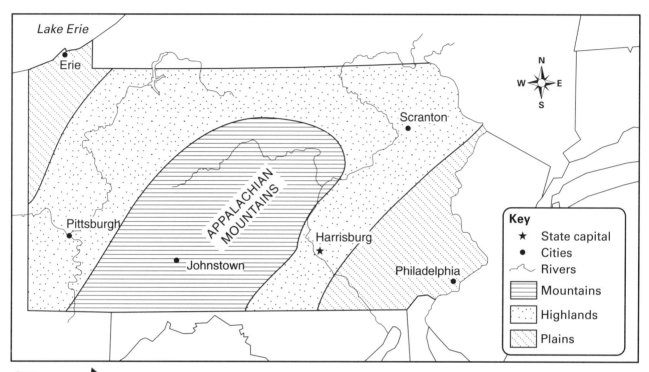

Directions Study the map of Pennsylvania. Use the map to answer the questions.

1. Which city is located in the highlands in northeast Pennsylvania?

2. Which city is located on the plains in southeast Pennsylvania?

3. Which city is located just west of the Appalachian Mountains?

4. Some relief maps label the highest mountain in the state. If this map showed the highest point in Pennsylvania, near which city would you expect it to be?

5. What is the most common landform in Pennsylvania: mountains, highlands, or plains?

Activity

Use the map above. Color the mountains red. Color the highlands orange. Color the plains green.

Name _____ Date _____

Elevation Maps

A landform map is sometimes called an **elevation map**. An elevation map shows the elevation, or height, of the land's surface. Study the elevation scale in the map below. It shows the elevation in feet and meters. The scale begins at 0 feet, which is also called **sea level**.

The map uses scattered dots to show elevation from 0 to 250 feet. Notice these dots along the coastal plain of Texas. A crisscross pattern is used to show elevation from 2,000 to 5,000 feet. Much of West Texas shows this pattern. Elevations above 5,000 feet are shown with dark vertical lines.

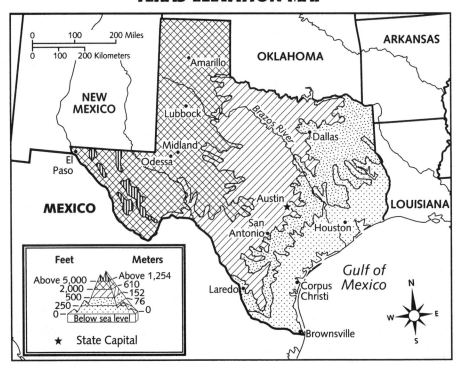

 Study the map and the elevation scale. Then, read the statements below. Write **T** if the statement is true. Write **F** if the statement is false.

_____ **1.** Most of Texas lies below sea level.

_____ **2.** The highest parts of Texas are located in the south.

_____ **3.** Lubbock is higher in elevation than Corpus Christi.

_____ **4.** Brownsville is lower in elevation than Dallas.

_____ **5.** Austin and San Antonio are found in the same range of elevation.

Activity

What is the highest point in your state? Do research to find out. Then, write a short report about what you learn.

Rivers

Sometimes landform maps are called **physical maps**. They show the physical features of a place. One important physical feature is a river. There are many major rivers in the United States. Rivers are useful for moving goods and people. Rivers are also useful for recreation or generating power. Some rivers serve as boundaries between states or nations.

Study the map key below. Find the symbol that stands for rivers. On this map, rivers are shown as wiggly lines. The names of the rivers are also included.

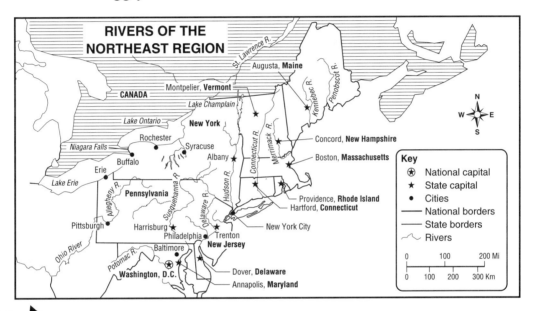

Directions Study the map. Use the map to answer the questions.

1. Through which city does the Allegheny River flow?

2. Which river flows through the national capital of the United States?

3. Which river flows near the state capital of Maine?

4. Which river can you use to travel from Albany, New York, to New York City?

5. A river forms the border between two states and then runs south through two other states. Is it the Allegheny River, Connecticut River, or Hudson River?

6. Name three states whose capital cities are next to rivers.

Name _____ Date _____

Resource and Product Maps

A **resource** is something that people use to make or produce things. Trees are a resource. People use them to make lumber and paper. Oil is another resource. People use it to make gasoline and plastics. There are many kinds of resources.

A **resource map** shows where these resources are located. Sometimes a resource map is called a **product map**. To use a resource or product map, first read the title of the map. It tells you what the map is about. Then, study the map key. The map key will have symbols that stand for resources or products. In the map below, a picture of an apple stands for fruit.

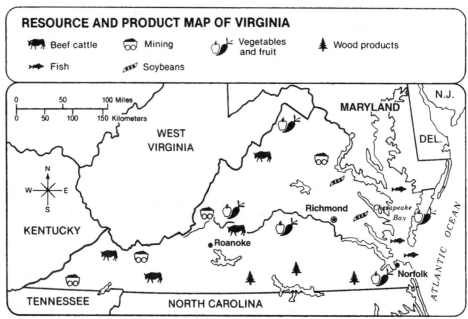

Directions Study the map and map key. Use the map to finish each sentence. Darken the circle by your answer choice.

1. Beef cattle are mainly found in the _____ part of Virginia.
   Ⓐ northeastern
   Ⓑ southeastern
   Ⓒ Chesapeake Bay
   Ⓓ western

2. The two main resources in the southwestern tip of Virginia are _____.
   Ⓐ vegetables and fruit
   Ⓑ mining and cattle
   Ⓒ soybeans and cattle
   Ⓓ wood products and fish

3. The main product in the Chesapeake Bay area of Virginia is _____.
   Ⓐ cattle
   Ⓑ wood products
   Ⓒ fish
   Ⓓ chickens

4. The product found north and east of Richmond is _____.
   Ⓐ cattle
   Ⓑ wood products
   Ⓒ soybeans
   Ⓓ vegetables and fruit

www.svschoolsupply.com
© Steck-Vaughn Company

Unit 6: How to Use a Resource Map
Map Skills 4, SV 6132-8

Name _____   Date _____

Land Use Maps

Sometimes, resource maps are known as **land use maps**. Land use maps tell how people use the land to produce things. Look at the map below. Study the map key. People in Oregon use their land in several ways.

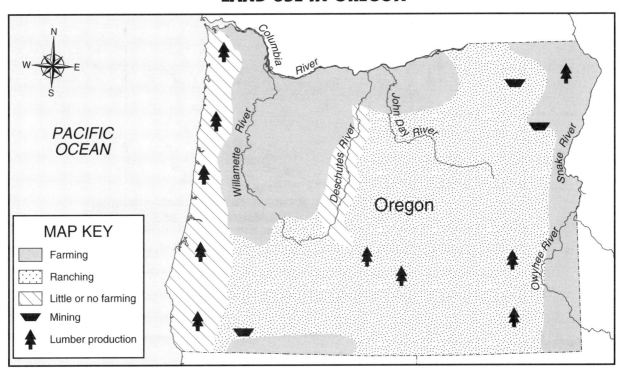

Directions → Study the map. Use the map to answer the questions.

1. What is the main way that people in Oregon use the land?

   _____

2. In what parts of Oregon do the people use the land for mining?

   _____

3. How is the land along the Pacific Coast used?

   _____

4. What are two ways the land is used in the center of Oregon?

   _____

5. How is the land along the Willamette, Columbia, and Snake rivers mainly used?

   _____

Name _____   Date _____

A Useful Map

Some maps show resources, products, and land use. These maps allow you to see all the uses people find for their land. Study the map below. It is of the imaginary state of East Albion. Notice all the ways people make use of the land.

EAST ALBION

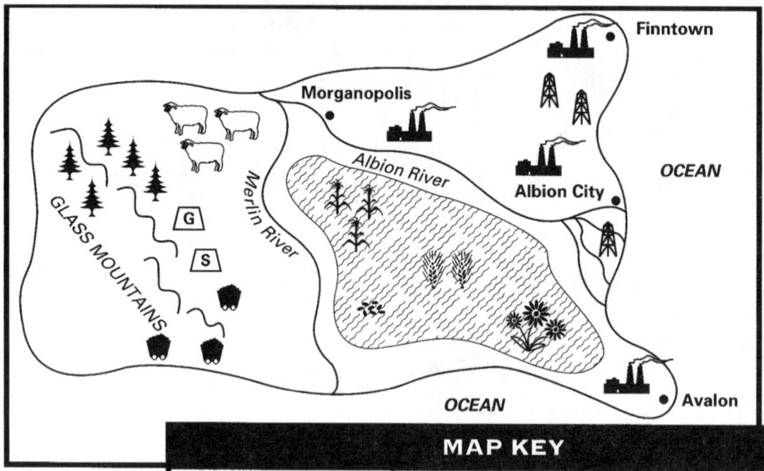

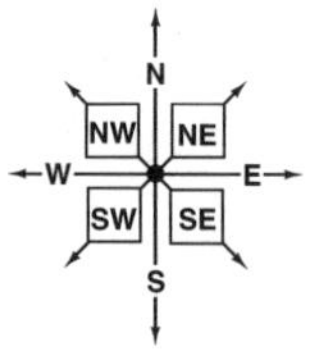

[Map showing East Albion with Finntown, Morganopolis, Albion City, and Avalon; Albion River and Merlin River; Glass Mountains; surrounded by OCEAN]

MAP KEY

LAND USE	LEADING RESOURCES	MAJOR CROPS
Forest	Coal	Corn
Sheep Ranches	G Gold	Wheat
Farming	S Silver	Soybeans
Factories	Oil	Flowers

[Compass rose showing N, NE, E, SE, S, SW, W, NW]

Directions ▶ Darken the circle by the correct answer to each question.

1. Where are the sheep ranches mainly found?
   - Ⓐ in the east
   - Ⓑ in the south
   - Ⓒ in the west
   - Ⓓ in the north

2. What three resources are found near the Glass Mountains?
   - Ⓐ gold, silver, and oil
   - Ⓑ coal, gold, and oil
   - Ⓒ coal, gold, and silver
   - Ⓓ coal, silver, and oil

3. Which part of East Albion is used more for forests?
   - Ⓐ northeast
   - Ⓑ southeast
   - Ⓒ southwest
   - Ⓓ northwest

4. How is the land between the Merlin River and the Albion River mainly used?
   - Ⓐ farming
   - Ⓑ factories
   - Ⓒ sheep ranches
   - Ⓓ forests

☀ Activity

Do research to find out about resources in your state. Then, draw a map of your state. Make a map key for your map. Draw pictures for the resources you have learned about. Draw the pictures on the map to show where the resources are found in your state.

Name _____ Date _____

Latitude

Some maps have imaginary lines that circle the Earth. Some lines circle the Earth from east to west. These are called **lines of latitude.** The lines help you to locate places on the Earth.

Each line of latitude is named with a number of **degrees**. The **Equator** is 0°. All the other lines of latitude are north or south of the Equator. They go up to 90° N (north) or down to 90° S (south).

Look at the map of Brazil. Do you see the lines that go east and west across the country? Those are lines of latitude. You can see that Brazil is between 5° N and 35° S. The Equator runs through the northern part of Brazil. Find the Equator on the map. Follow it with your finger from east to west.

Sometimes a place falls between two lines of latitude. Then, you must estimate its line of latitude. For example, look at the city of Salvador on the map of Brazil. Salvador is between 10° S and 15° S. You could estimate that its line of latitude is about 13° S.

 Study the map. Use the map to answer the questions. Darken the circle by your answer choice.

1. Which city is located on the Equator?
   Ⓐ Recife   Ⓑ Macapá   Ⓒ Manaus

2. What is the latitude of Rio de Janeiro?
   Ⓐ 10° N   Ⓑ 18° S   Ⓒ 22° S

3. What is the latitude of Fortaleza?
   Ⓐ 4° N   Ⓑ 4° S   Ⓒ 6° S

4. Which city has a latitude of about 16° S?
   Ⓐ Fortaleza   Ⓑ Manaus   Ⓒ Brasília

5. Which city has a latitude of about 3° S?
   Ⓐ Manaus   Ⓑ Recife   Ⓒ São Paulo

6. How many large cities in Brazil are south of 10° S?
   Ⓐ 0   Ⓑ 5   Ⓒ 10

Name _____   Date _____

Longitude

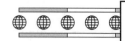 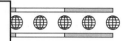

You have learned about lines of latitude. There are also imaginary lines that circle the Earth from north to south. These are called **lines of longitude**. These lines also help you to locate places on the Earth.

Each line of longitude is named with a number of degrees. The **Prime Meridian** is 0°. All the other lines of longitude are east or west of the Prime Meridian. They go up to about 179° E (east) and 179° W (west). There is only one 180° line of longitude. Like the Prime Meridian, it does not have a direction.

Look at the map of Portugal and Spain. Do you see the lines that go north and south across the countries? Those are lines of longitude. You can see that Portugal and Spain are between 4° E and 10° W. The Prime Meridian runs through the eastern part of Spain. Find the Prime Meridian on the map. Follow it with your finger from north to south.

If a place falls between two lines of longitude, you must estimate its line of longitude. Look at Lisbon on the map. It falls between 8° W and 10° W. You could estimate that its line of longitude is 9° W.

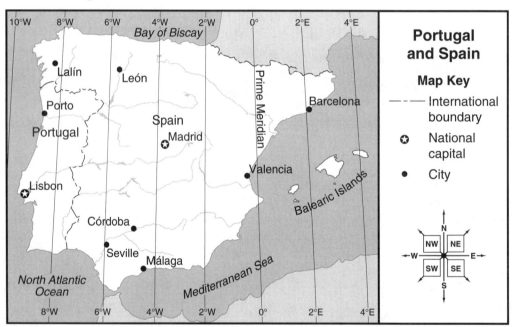

Directions ➔ Study the map. Use the map to answer the questions.

1. Find the national capital of Spain. What is its name? _____

   What line of longitude is it near? _____

2. Which city in Spain is nearer to the Prime Meridian, Valencia or Barcelona?

   _____

3. Find the national capital of Portugal. What is its name? _____

   Estimate its line of longitude. _____

Name _____   Date _____

Latitude and Longitude

NIGERIA

You have learned that lines of latitude run east and west around the Earth. Lines of longitude run north and south. Together, the lines of latitude and longitude form a grid on maps and globes. By using lines of latitude and longitude, you can find any place on Earth.

Look at the map of Nigeria. Kano and Maiduguri are both near 12° N latitude. Of course, they are not in the same place. By adding the longitude of each place, you can show exactly where it is. The latitude of a place is written first. The longitude is written next to it. Look at Kano. We can say that Kano is near 12° N, 9° E.

Directions → Darken the circle by the correct answer to each question.

1. What city is the national capital of Nigeria?
   - Ⓐ Sokoto
   - Ⓑ Abuja
   - Ⓒ Kano
   - Ⓓ Port Harcourt

2. Which city in Nigeria is located nearest to 4° N?
   - Ⓐ Sokoto
   - Ⓑ Abuja
   - Ⓒ Lagos
   - Ⓓ Port Harcourt

3. Which city has a latitude and longitude of about 6° N, 3° E?
   - Ⓐ Maiduguri
   - Ⓑ Sokoto
   - Ⓒ Lagos
   - Ⓓ Port Harcourt

4. The latitude and longitude of Abuja is about _____.
   - Ⓐ 9° N, 9° E
   - Ⓑ 9° N, 7° E
   - Ⓒ 12° N, 8° E
   - Ⓓ 5° N, 7° E

Activity

Find where you live on a map of the United States or of your state. If you cannot find your town, find a city near it. Find the lines of latitude and longitude near where you live. Write them down.

Name _____ Date _____

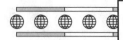 # Can You Find It?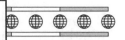

You have learned about lines of latitude and longitude. Now let's use what you have learned. Look at the map of the United States. The numbers for the lines of latitude are on the left. The numbers for the lines of longitude are at the top. To help you find a place, move your finger along the lines.

THE UNITED STATES

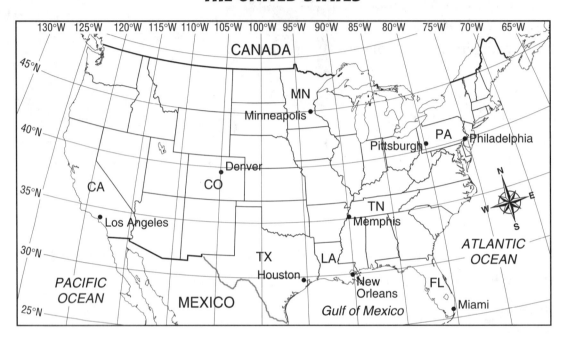

Directions ➔ Study the map. Use the map to answer the questions.

1. Which city is located at 40° N, 75° W?

   _____

2. Which two cities are located along 80° W?

   _____

3. Which city is located at 34° N, 118° W?

   _____

4. Which city is at the same latitude as New Orleans?

   _____

5. Draw a mountain beside the city located at 40° N, 105° W.

 Activity

Get your teacher or an adult to help you with this activity. Write the latitude and longitude for your town. Then, go to this address on the Internet:
http://www.fourmilab.ch/cgi-bin/uncgi/Earth
This is the web address for EarthCam. At this site you can type in your latitude and longitude. Then, you can see your area from a camera in space!

Name _____ Date _____

Globes

The Earth is shaped like a round ball. A **globe** is a model of the Earth. So a globe is shaped like a round ball, too. A globe shows the large, main pieces of land, called **continents**. There are seven continents. A globe shows the main bodies of water, called **oceans**. There are four oceans.

Globes have a compass rose to help you to find directions. Some globes have a map key and distance scale. Most globes also have lines of latitude and longitude.

Do you have a globe? If you do, look at it. If not, look at the map below. At the top of the globe is the North Pole. At the bottom of the globe is the South Pole. In the middle of the globe is the Equator.

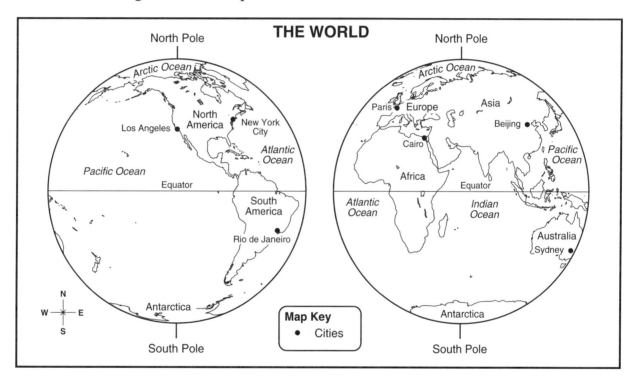

Directions Study a globe or the map. Use it to answer the questions.

1. What are the names of the seven continents? _____
_____   _____   _____
_____   _____   _____

2. What are the names of the four oceans?
_____   _____
_____   _____

3. Which continent do you live on? _____

www.svschoolsupply.com
© Steck-Vaughn Company

Unit 8: How to Use Globes and World Maps
Map Skills 4, SV 6132-8

Name _____   Date _____

World Maps

World maps are not round like globes. World maps are flat. On a world map, Australia is just shown as being east of South America. On a globe, you can see that Australia is on the opposite side of the world from South America. Globes are good for showing where places really are on the Earth. World maps are good for helping you to know what all those places are.

Look at the world map. You can see all the continents and oceans. On a globe, you would only see half of those things. You would have to turn the globe to see the rest.

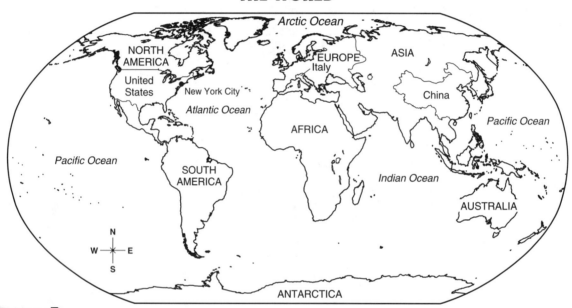

Directions Study the map. Use the map to answer the questions. Darken the circle by your answer choice.

1. Which continent is farthest south?
   - Ⓐ South America
   - Ⓑ Australia
   - Ⓒ Africa
   - Ⓓ Antarctica

2. To travel directly from Europe to Australia, which direction must you go?
   - Ⓐ northwest
   - Ⓑ southeast
   - Ⓒ northeast
   - Ⓓ southwest

3. Which ocean is the most northern?
   - Ⓐ Arctic
   - Ⓑ Atlantic
   - Ⓒ Pacific
   - Ⓓ Indian

4. To travel directly from Europe to the United States, which ocean must you cross?
   - Ⓐ Arctic
   - Ⓑ Atlantic
   - Ⓒ Pacific
   - Ⓓ Indian

5. To travel directly from Australia to Africa, which ocean must you cross?
   - Ⓐ Arctic
   - Ⓑ Atlantic
   - Ⓒ Pacific
   - Ⓓ Indian

6. Which ocean touches both South America and Asia?
   - Ⓐ Arctic
   - Ⓑ Atlantic
   - Ⓒ Pacific
   - Ⓓ Indian

Name _____   Date _____

Hemispheres

A **hemisphere** is a half of a globe. The Earth has four hemispheres. This sounds like the Earth has four halves. It does, in a way. The Equator splits the Earth into two halves, the Northern Hemisphere and the Southern Hemisphere. The Prime Meridian splits the Earth into two hemispheres, too. Those two are the Western Hemisphere and the Eastern Hemisphere. So the Earth does have four halves!

Most places are in two hemispheres. North America is in the Northern Hemisphere. It is also in the Western Hemisphere. Australia is in the Southern Hemisphere. It is also in the Eastern Hemisphere. Africa has places in all four hemispheres!

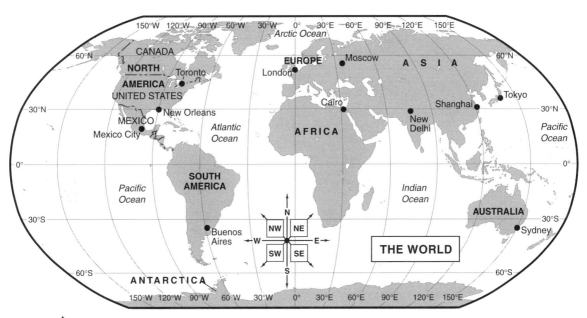

Directions Study the map. Use the map to answer the questions.

1. In which three hemispheres is South America located?

   _____
   _____
   _____

2. In which three hemispheres is Europe located?

   _____
   _____
   _____

☼ Activity

Use the map above. Color the Equator red. It is the 0° line of latitude. It runs east and west. Color the Prime Meridian black. It is the 0° line of longitude. It runs north and south. Color the Northern Hemisphere yellow. Color the Southern Hemisphere brown. Color the Western Hemisphere green. Color the Eastern Hemisphere blue. Isn't the world a colorful place?

www.svschoolsupply.com
© Steck-Vaughn Company

39

Unit 8: How to Use Globes and World Maps
Map Skills 4, SV 6132-8

Name _____ Date _____

The Eastern Hemisphere

The Eastern Hemisphere is half of the Earth. The Eastern Hemisphere starts at the Prime Meridian. It goes eastward to 180°. It covers all of that part of the Earth from the North Pole to the South Pole. Do you live in the Eastern Hemisphere?

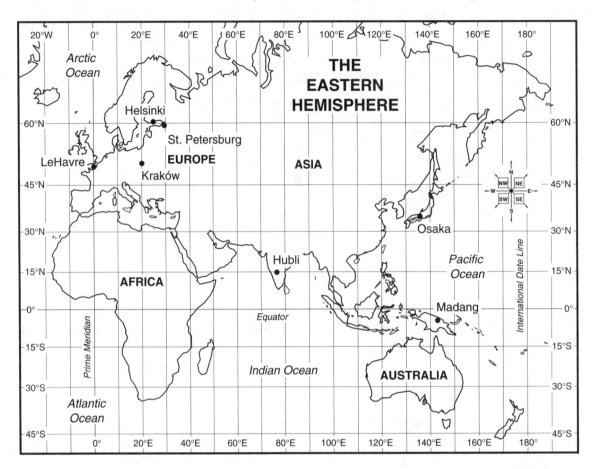

Directions Study the map. Use the map to answer the questions.

1. Which ocean touches Africa, Asia, and Australia?

2. Which city is located on the Prime Meridian?

3. Which city is located at 37° N, 135° E?

4. Do the Equator and Prime Meridian meet on land or on water?

5. At what latitude and longitude do the Equator and Prime Meridian meet?

6. What are the latitude and longitude of Hubli?

www.svschoolsupply.com
© Steck-Vaughn Company

Unit 8: How to Use Globes and World Maps
Map Skills 4, SV 6132-8

Name _____ Date _____

North America

The United States is on the continent of North America. North America is in the Northern Hemisphere. It is also in the Western Hemisphere. Do you live in North America?

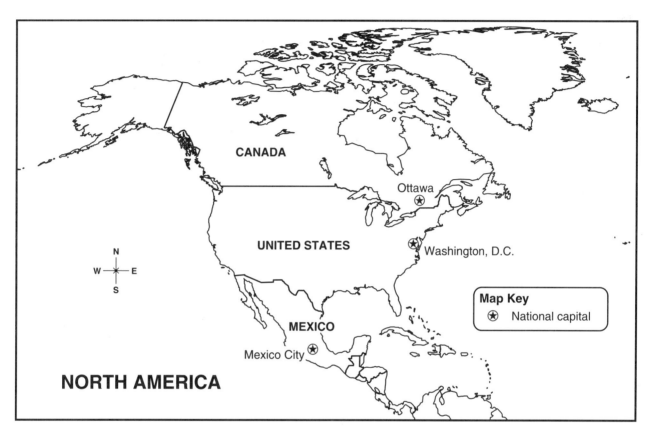

Directions Study the map. Use the map to answer the questions.

1. Which three countries make up most of North America?
   _____
   _____
   _____

2. Which country in North America is the farthest north?
   _____

3. Which city is the national capital of Mexico?
   _____

4. Which city is the national capital of Canada?
   _____

5. If you traveled from Mexico City to Washington, D.C., which direction would you go?
   _____

Name _____ Date _____

Historical Maps

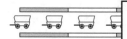 

The world has changed over time. The names and borders of some states and countries have changed. Routes have changed, too. Some maps tell about places in the past. These maps are called **historical maps**. They help us to learn how the world used to be.

Look at the map below. It shows the middle part of the United States in the 1870s. Is that part the same way now? Look at the shaded states just west of Minnesota. In the 1870s, that area was known as Dakota Territory. Now, that area makes up two states, North Dakota and South Dakota.

Directions ➡ Study the map. Use the map to answer the questions.

1. Which river forms the southern boundary of Indiana?

   _____

2. Which river forms the boundary between Iowa and Illinois?

   _____

3. Which river flows by Fort Randall in Dakota Territory?

   _____

4. In which state does the Mississippi River begin?

   _____

 Activity

Imagine that you are traveling by raft down the Mississippi River in the mid-1800s. What would see on your journey? Write a story about your adventures. Include a drawing.

www.svschoolsupply.com
© Steck-Vaughn Company

Unit 9: How to Use a Historical Map
Map Skills 4, SV 6132-8

Name _____ Date _____

Trails to the Past

Routes have changed over time. In the 1800s, people used trails to head west. They found many adventures along the Oregon Trail or the California Trail. The Oregon Trail saw its heaviest use in the 1840s and 1850s. By 1870, the railroad stretched from coast to coast. Pioneers no longer used the Oregon Trail. By that time, though, thousands of Americans had traveled in wagon trains along the Oregon Trail.

Look at the map of the Oregon Trail. Notice the small map in the upper right corner of the map. This is called an **inset map**. It shows the part of the country that the larger map tells about.

Directions → Study the map. Use the map to answer the questions.

1. About how long was the Oregon Trail?

   _____

2. In which cities did the Oregon Trail begin and end?

   _____
   _____

3. Which river did the Oregon Trail follow east of the Rocky Mountains?

   _____

4. Which river did the Oregon Trail follow west of the Rocky Mountains?

   _____

Activity

Imagine that you and your family traveled west on the Oregon Trail. What kinds of problems or adventures would you have along the way? Write a short story about what might happen to you on the trip. Draw a picture of your wagon train.

Name _____ Date _____

Time Zone Maps

When the Sun shines on the Earth, it is daytime on that side of the Earth. But on the other side, it is nighttime. Because the Earth is so big, it is divided with imaginary lines to show different **time zones**.

The Earth has 24 time zones. All the people in the same time zone set their clocks to the same time. Imagine that the time zones are numbered 1 to 24. The number 1 zone is at the International Date Line, or 180° longitude. As the number of the time zone increases by 1, the time becomes an hour later.

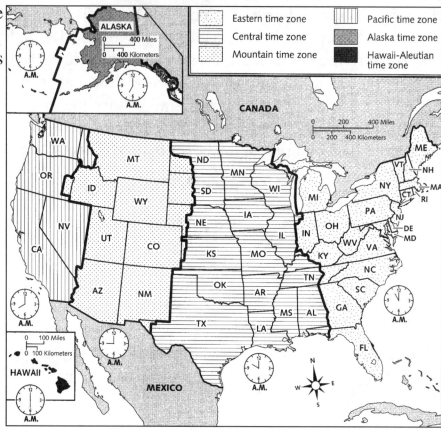

The different time zones allow people around the world to experience a common event at about the same clock time. For example, most everyone in the world has sunrise at about 6:00 in the morning.

The United States has six time zones. As you move east from one time zone to another, the time gets an hour later.

Directions > Study the map. Use the map to answer the questions.

1. In which time zone do you live? _____
2. When it is 4:00 P.M. in New Mexico, what time is it in Maine? _____
3. When it is 11:00 A.M. in Louisiana, what time is it in California? _____
4. You drive from Alabama to Georgia. Will you have to set your clock forward an hour or backward an hour when you arrive? _____

Activity

Write a story about going backward or forward in time. To what time would you go? Include a drawing of your time machine.

Name _____ Date _____

Population Maps

A **population map** shows how many people live in a certain area. This kind of map shows the density of population. Density of population means how many people live in a certain area. Usually, that area is a square mile. Look at the map below. It shows the population of a region in the United States. Notice the map key. It shows the density scale, or how many persons per square mile.

Study the different kinds of shading. Small, close dots mean over 500 persons per square mile. Where would these places be on the map? Look at the cities around the lakes, such as Chicago, Detroit, and Cleveland. These cities all have an average of over 500 people per square mile. Notice the wavy lines that mean 0 to 4 people per square mile. Where are these on the map?

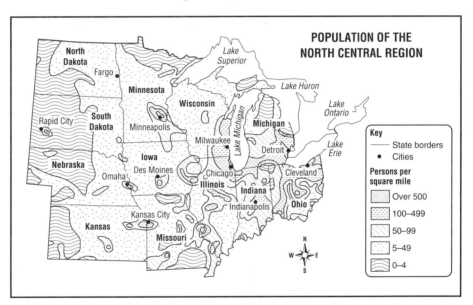

Directions: Study the map. Use the map to answer the questions.

1. How many people per square mile are there around Fargo, North Dakota?

   _____

2. About how many people per square mile live in western Kansas?

   _____

3. Which state seems to have the least dense population?

   _____

4. Which state seems to have the most dense population?

   _____

Activity

How many people live in your town? How many square miles are in your town? Do research to find out. Then, divide the number of people by the square miles. The answer will be the density of population in your town.

www.svschoolsupply.com
© Steck-Vaughn Company

Unit 10: How to Use Special Purpose Maps
Map Skills 4, SV 6132-8

Name _____   Date _____

Precipitation Maps

Precipitation is the moisture that falls from the sky. It can be rain, snow, sleet, or hail. A **precipitation map** shows how much precipitation falls in a place. The map usually shows the inches of precipitation per year.

Look at the map. It shows the amount of precipitation in the Southwest Region. Study the map key. Notice the shading. For example, horizontal lines mean less than 10 inches of precipitation a year. Diagonal dots mean 40 to 60 inches a year. Some places on the map are very wet, and others are very dry.

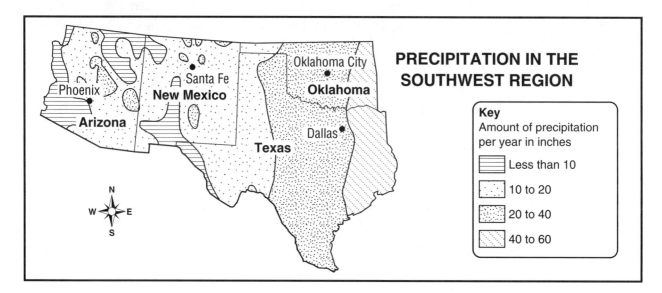

Directions → Study the map. Use the map to answer the questions.

1. About how much precipitation per year falls around Oklahoma City?

2. About how much precipitation per year falls around Santa Fe?

3. Which part of Texas gets the most precipitation?

4. According to the map, which state probably gets more precipitation per year, Oklahoma or New Mexico?

5. Do you think farming would be an important job in Arizona? Explain.

www.svschoolsupply.com
© Steck-Vaughn Company

Unit 10: How to Use Special Purpose Maps
Map Skills 4, SV 6132-8

Name _____ Date _____

Movement Maps

A **movement map** shows how things move. It can show how people or animals migrate from one place to another. It can show how products move from their source to a factory. It can show how the ocean waters or upper-level winds move.

Study the map below. The Hawaiian islands were once bare volcanic rocks. All of the plants, animals, and birds came there from other parts of the world. Many were brought to Hawaii by early explorers and settlers. Look at the arrows. Each arrow has a letter on it. Look at the list of letters in the map key. Each letter stands for an animal or plant. Letter **B** stands for bats. Find the arrow with the letter **B** on it. Bats came to Hawaii from which country?

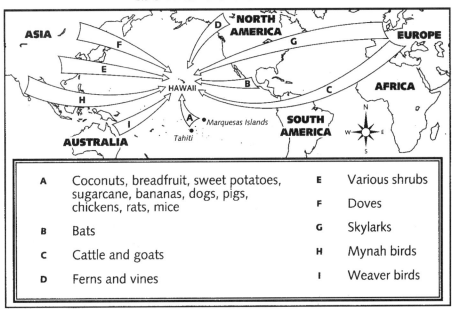

MOVEMENT TO HAWAII

A	Coconuts, breadfruit, sweet potatoes, sugarcane, bananas, dogs, pigs, chickens, rats, mice	E	Various shrubs
		F	Doves
B	Bats	G	Skylarks
C	Cattle and goats	H	Mynah birds
D	Ferns and vines	I	Weaver birds

Directions ➤ Study the map. Use the map to answer the questions.

1. Which animals traveled east from Asia to reach Hawaii?

   _____

2. Which plants came to Hawaii from North America?

   _____

3. From which continent did cattle and goats come to Hawaii?

   _____

4. Which direction did dogs and rats travel to reach Hawaii?

   _____

Activity

Draw a map of your school. Use arrows to show your movement around the school during the day. Write a letter on each arrow. Include a list of letters and a description of what each letter means.

www.svschoolsupply.com
© Steck-Vaughn Company

47

Unit 10: How to Use Special Purpose Maps
Map Skills 4, SV 6132-8

Answer Key for Map Skills, Grade 4

Pretest, page 3
1. C, 2. A, 3. C, 4. D, 5. A, 6. C

Posttest, page 4
1. Rome, 2. Berlin, 3. northwest, 4. Oslo, Norway, 5. London, England, 6. 60° N, 25° E

page 10
1. northeast, 2. no

page 11
1. D, 2. B, 3. B, 4. C

page 12
1. Old Town, 2. school, 3. railroad, 4. bike path

page 13
1. B-1, 2. hospital, 3. A-3 and B-3, 4. C-1, D-1, and D-2

page 14
1. B-2, 2. D-1; northeast, 3. D-3, 4. B-2

page 15
1. Four Cities, 2. west, 3. northeast, 4. southwest, 5. northeast, 6. northwest, 7. north and south, 8. east and west

page 16
1. about 1 mile, 2. about 2 miles

page 17
1. $\frac{1}{2}$, 2. 150, 3. 240, 4. $1\frac{1}{2}$, 5. 450, 6. Check students' maps.

page 18
1. Independence: A-4; Park Headquarters: C-2; Mineral King: C-3; Lodgepole: B-2; Cedar Grove: A-2; Mt. Whitney: B-4, 2. C-2, B-2, C-3, 3. northeast, 23, 4. northwest, 17, 5. south, 7, 6. west, 25

page 20
1. Answers will vary: Go west on Market Street, then south on Fifth Street., 2. about 1,700 feet, 3. Market Street, Chestnut Street, Walnut Street, Spruce Street, 4. at the stable

page 21
1. State Highway 41, 2. Interstate Highway 3, U.S. Highway 13, State Highway 41, 3. 14 miles, 4. south on Interstate Highway 3 to Russell, then west on State Highway 41 to Schnell

page 22
1. 236, 2. 221, 3. 157, 4. 353

page 23
1. Hartford, 2. northeast, 3. Long Island Sound, 4. Answers will vary: Go northeast on Interstate Highway 95 to New Haven, then north on Interstate Highway 91 to Hartford.

page 24
1. Quito, 2. Peru, 3. 150 miles; 250 kilometers, 4. Pacific Ocean, 5. 100 miles; southeast, Activity: Lima, Peru; Bogota, Colombia

page 25
1. Columbus, 2. Washington, D.C., 3. Cheyenne, 4. Atlanta, 5. Canada, 6. Mexico, 7. southeast, 8. east, 9. Nashville, Tennessee, 10. Denver, Colorado

page 26
1. plains, 2. mountains, 3. hills, 4. mountains

page 27
1. Scranton, 2. Philadelphia, 3. Pittsburgh, 4. Johnstown, 5. Answers may vary: highlands

page 28
1. F, 2. F, 3. T, 4. T, 5. T

page 29
1. Pittsburgh, Pennsylvania, 2. Potomac River, 3. Kennebec River, 4. Hudson River, 5. Connecticut River, 6. Students should name three of these: Maine, New Hampshire, Connecticut, New Jersey, New York, Pennsylvania.

page 30
1. D, 2. B, 3. C, 4. C

page 31
1. ranching, 2. northeast, southwest, 3. lumber production, 4. ranching, lumber production, 5. farming

page 32
1. D, 2. C, 3. D, 4. A

page 33
1. B, 2. C, 3. B, 4. C, 5. A, 6. B

page 34
1. Madrid; 4° W, 2. Valencia, 3. Lisbon; 9° W

page 35
1. B, 2. D, 3. C, 4. B

page 36
1. Philadelphia, 2. Pittsburgh, Miami, 3. Los Angeles, 4. Houston, 5. Students should draw a mountain near Denver.

page 37
1. Asia, Africa, Australia, Antarctica, Europe, North America, South America, 2. Arctic Ocean, Atlantic Ocean, Indian Ocean, Pacific Ocean, 3. North America

page 38
1. D, 2. B, 3. A, 4. B, 5. D, 6. C

page 39
1. Northern Hemisphere, Southern Hemisphere, Western Hemisphere, 2. Northern Hemisphere, Eastern Hemisphere, Western Hemisphere

page 40
1. Indian Ocean, 2. LeHavre, 3. Osaka, 4. water, 5. 0°, 0°, 6. 15° N, 78° E

page 41
1. United States, Canada, Mexico, 2. Canada, 3. Mexico City, 4. Ottawa, 5. northeast

page 42
1. Ohio River, 2. Mississippi River, 3. Missouri River, 4. Minnesota

page 43
1. about 1,800 miles, 2. Independence, Missouri, to Portland, Oregon Territory, 3. North Platte River, 4. Snake River

page 44
1. Answers will vary. 2. 6:00 P.M., 3. 9:00 A.M., 4. forward

page 45
1. 5 to 49, 2. 0 to 4, 3. Answers may vary: South Dakota, 4. Answers may vary: Ohio

page 46
1. 20 to 40 inches, 2. 10 to 20 inches, 3. east, 4. Oklahoma, 5. No, because most farming takes a lot of precipitation, and Arizona does not get much; much of Arizona is desert.

page 47
1. doves and mynah birds, 2. ferns and vines, 3. Europe, 4. north